CURIOUS FEATURES OF EXTRAORDINARY CREATURES

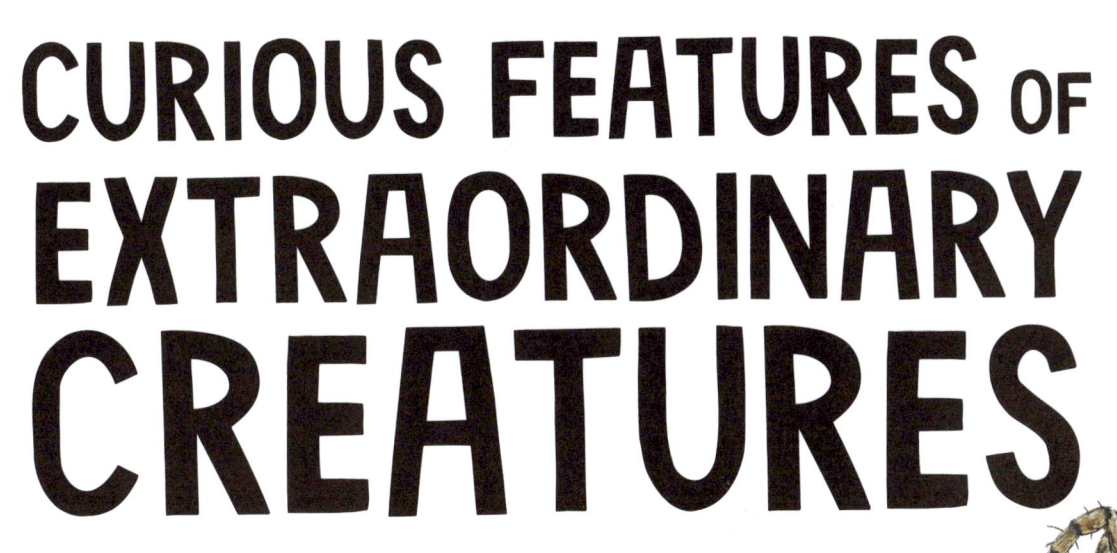

CURIOUS FEATURES OF EXTRAORDINARY CREATURES

MACMILLAN CHILDREN'S BOOKS

For Sam. With huge thanks to my family, whose
ongoing support has been invaluable - FF

For Tristan, a precious and curious creature - CdlB

First published 2024 by Macmillan Children's Books
an imprint of Pan Macmillan
The Smithson, 6 Briset Street, London, EC1M 5NR
Associated companies throughout the world
www.panmacmillan.com

EU representative: Macmillan Publishers Ireland Ltd, 1st Floor,
The Liffey Trust Centre, 117-126 Sheriff Street Upper, Dublin 1, D01 YC43

ISBN 978-0-7534-4978-3

Illustrator and concept by: Fiona Fogg
Editor and consultant: Camilla de la Bedoyere
Senior designer: Lisa Sodeau
Senior commissioning editor: Elizabeth Yeates

9 8 7 6 5 4 3 2 1

A CIP catalogue record for this book is available from the British Library.

Printed in China

MIX
Paper | Supporting
responsible forestry
FSC® C116313

Contents

Introduction

Welcome to the wonderful world of curious creatures and prepare to be astonished. There are billions of animals alive on our precious planet today, and they are all special in their own way, but some types are more peculiar than others!

Have you ever heard of an animal with a tongue as long as its body, a beautiful spider that loves to dance or a clever toad that grows babies in its skin? Did you know there are small blue dragons that float on the waves, slimy scavengers that tie themselves in knots and nose-picking lemurs that eat grubs... and snot?

This book is a colourful treasury of some of the most incredible and
extraordinary creatures of the animal kingdom. Read the stories of
how they live their lives and what makes them so unique.

As you turn the pages, think about the way that each animal's curious qualities help
it to stay alive, find food or have young. If you spend time in the natural world, you
will soon discover that every creature on Earth has its own special qualities.
That is why it is so important we treasure all of them.

Greater Bird Of Paradise

A bird of beauty and grace

Birds of paradise are amongst the most beautiful animals on Earth. They have stunning feathers in jewel-like colours and perform dazzling dance displays.

One by one...

... **males arrive**.

They **begin their dance** by stretching.

Then spread out their lovely **tail feathers**.

It's time to **quiver!**

And **shimmy and shake**.

And **run up and down** the branch.

Success! The female is impressed.

She moves closer to the male **she likes most**.

He woos her by **gently pecking** her feathers.

The long thread-like feathers in a bird of paradise's plumage are called wires. Long, soft feathers are called plumes.

Male greater birds of paradise sing as they perform stunning dances. They are not very musical birds – their songs are a mix of 'wonk', 'eee-ak', 'wa-wa-wa' and 'baa'!

Birds often dance, sing and show off their fine feathers to impress their mates. This is known as courtship. Males gather in courtship groups, called leks.

While males are resplendent in their fine feathers, females are a dull purple-brown.

Fact File

Latin name: **Paradisaea apoda**
Origin: **New Guinea**
Habitat: **Tropical forests**
Size: **35-43 cm**
Lifespan: **Unknown**
Diet: **Fruit and insects**
Predators: **Humans, birds of prey and snakes**
Curiosity: **Incredible courtship dance**

Aye-Aye

Spindly-fingered nose-picker

Rarely seen, aye-ayes leave their leafy nests at night to climb through the tallest branches of forest trees, hunting for food. These dark-furred, scraggly-looking mammals are lemurs – animals closely related to monkeys and apes.

Like other lemurs, aye-ayes live on the island of Madagascar.

Unusually, their front teeth keep growing, so they never wear down.

This mysterious animal is now endangered, as its forest habitats have been cut down.

Their long middle-fingers are useful tools. They are the perfect size and shape for digging grubs (insect larvae) out of trees.

An aye-aye also uses its middle finger to go up its nose and scoop up a dollop of snot. The aye-aye then pulls its finger out of its nose and swallows the snot!

Fact File

Latin name: **Daubentonia madagascariensis**
Origin: **Madagascar (an island country off the east coast of Africa)**
Habitat: **Forests**
Size: **Body length 40 cm**
Lifespan: **Up to 20 years**
Diet: **Insects and plants**
Predators: **Unknown**
Curiosity: **Fingers for nose-picking and grub-grabbing**

The aye aye uses its **bony finger** to tap on wood…

… and listens for the sound of **juicy beetle grubs** living inside.

Using its **tough teeth**, the aye-aye gnaws a hole into the wood.

It pokes its finger into the hole and **twists and turns** it, until it touches the grub.

The aye-aye spikes the grub with its **hook-like nail…**

… and pulls it out.

Axolotl

The tadpole that never grows up

An amphibian is a type of animal that leads two lives. Think of a frog: it begins its life as a tadpole, swimming in water, but grows into a four-legged adult that leaps about on land. An axolotl, however, is no ordinary amphibian. It is one of the world's weirdest animals because it never truly grows up.

Amphibians go through a big body change, called a metamorphosis, when they grow into adults. Axolotls spend their whole lives as tadpoles instead. This is called neoteny.

Most tadpoles live in water and breathe using feathery gills on either side of their head.

Like other salamanders, axolotls can **grow new body parts**. This amazing ability is called **regeneration**.

If an axolotl is attacked, it can **lose a limb and escape**.

The wound **quickly heals** and…

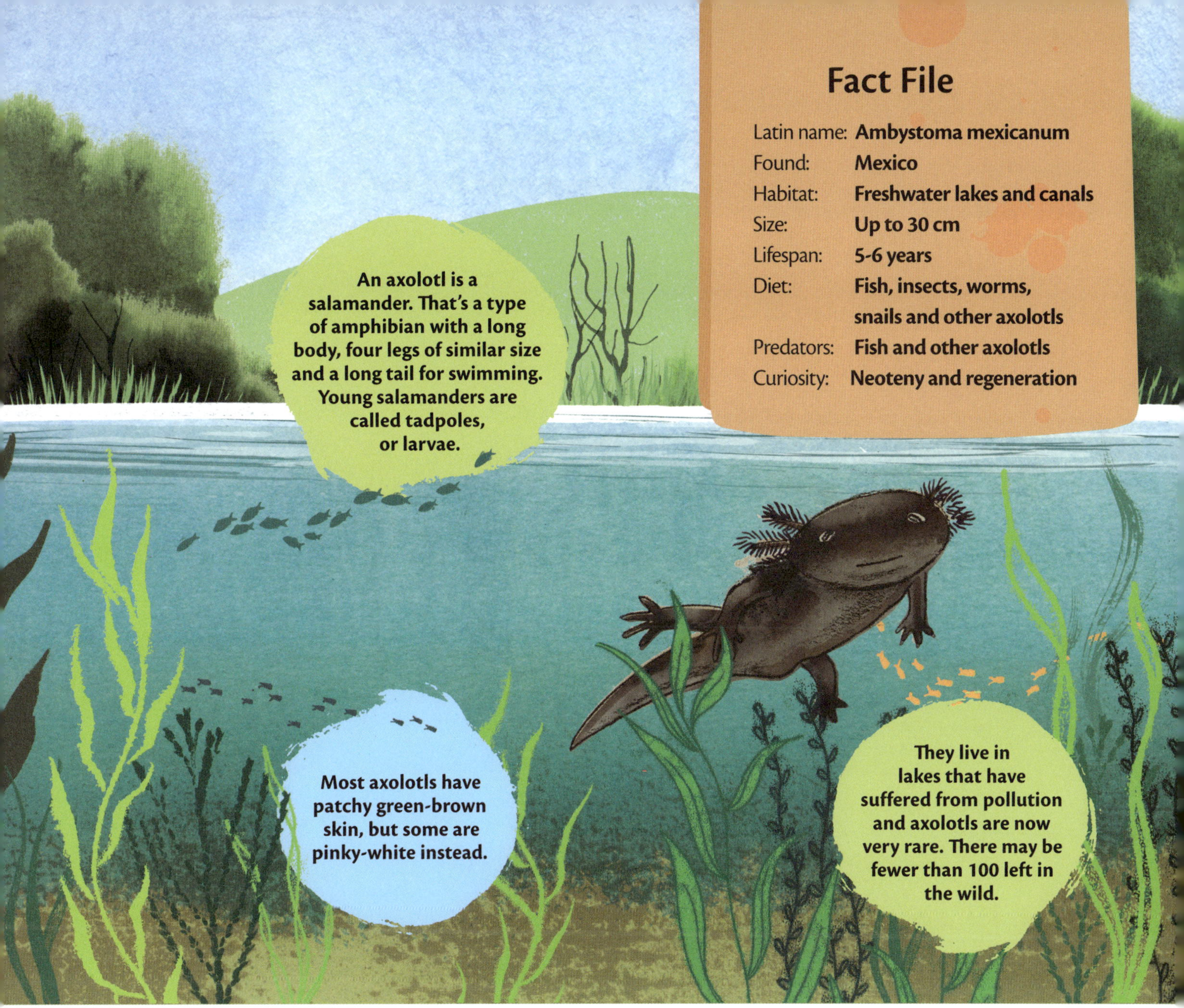

Fact File

Latin name:	**Ambystoma mexicanum**
Found:	**Mexico**
Habitat:	**Freshwater lakes and canals**
Size:	**Up to 30 cm**
Lifespan:	**5-6 years**
Diet:	**Fish, insects, worms, snails and other axolotls**
Predators:	**Fish and other axolotls**
Curiosity:	**Neoteny and regeneration**

An axolotl is a salamander. That's a type of amphibian with a long body, four legs of similar size and a long tail for swimming. Young salamanders are called tadpoles, or larvae.

Most axolotls have patchy green-brown skin, but some are pinky-white instead.

They live in lakes that have suffered from pollution and axolotls are now very rare. There may be fewer than 100 left in the wild.

... a **new limb** starts to grow.

It grows...

... and grows...

... until...

... the limb has **regenerated**, good as new!

Bobbit Worm

A speedy, freaky worm with lethal weapons

Bobbit worms lurk, buried in the soft seabed, waiting for a chance to ambush a passing fish. When they sense movement nearby, bobbit worms spring into action with some surprising skills.

Five antennae – sense organs – are on the worm's head. They detect prey.

Bobbits are colourful worms. They shimmer like gold and red shiny metal.

A special mouthpart, called a pharynx, has razor-sharp jaws.

Thousands of bristles along the worm's body can deliver painful stings to humans.

The worm reacts with lightning-quick speed and can slice a fish in two.

Fact File

Latin name:	**Eunice aphroditois**
Origin:	**Worldwide**
Habitat:	**Warm oceans**
Size:	**Up to 3 m long**
Lifespan:	**3-5 years**
Diet:	**Fish, worms and algae**
Predators:	**Unknown**
Curiosity:	**Snappy jaws**

Lurking in the sand, a bobbit worm is waiting for passing prey.

There's movement.

The worm's **super-sensitive antennae** detect a fish.

SNAP!

There's no escaping the worm's **strong hold**.

The fish is dragged back into the **worm's burrow**.

Anglerfish

A fish that makes light to trick its prey

Deep in the inky-black ocean, there lurks a weird fish with toothy jaws. This is a female deep-sea anglerfish and she has a lazy, but brilliant, way to tempt other fish into her cavernous mouth.

Fact File

Latin name: **Caulophryne jordani**
Found: **Worldwide**
Habitat: **Deep oceans**
Size: **Up to 20 cm**
Lifespan: **Unknown**
Diet: **Fish**
Predators: **Unknown**
Curiosity: **Light-lures and tiny males**

A bony rod grows out of an anglerfish's head and dangles a lure in front of her mouth, like a fishing rod with bait.

Bacteria grow on the tip of the lure. They produce light, which make the lure glow in the dark.

A passing fish sees the light and thinks it means food. The fish swims up to the lure... and the anglerfish gobbles it up. She has a stretchy stomach, so she can swallow fish that are bigger than she is!

A fanfin anglerfish uses her long fin rays to sense movement in the water.

Male anglerfish are **tiny** and look different to females.

When a male finds a female, he **latches on to** her...

... **biting her flesh** with his tiny teeth. His **body fuses** with hers.

Like a parasite, the male **absorbs food** from his mate's bloodstream.

Three-Toed Sloth

A green tree-hugger with dung-munching friends

Sloths are famous for being sluggish mammals, which spend most of their lives in trees where they hang motionless for hours at a time. Their greenish fur is key to a special sloth secret.

This is a brown-throated three-toed sloth. Like other sloths, it moves slowly on land, but it is a strong, speedy swimmer.

Sloths look and smell like trees. They are almost impossible to see when they are hanging in the forest canopy. That helps them stay safe when predators are about.

Silvery-grey moths spend their whole lives nestled in a sloth's fur. When they die, their bodies help feed the fungi growing on the sloth.

Fungi and tiny green plants, called algae, grow in the sloth's fur, turning it green and camouflaging the sloth.

A sloth's fur is crawling with insects – 980 bugs have been found living on one sloth!

Fact File

Latin name:	**Bradypus variegatus**
Origin:	**Central and South America**
Habitat:	**Tropical forests and swamps**
Size:	**60 cm long**
Lifespan:	**30-40 years**
Diet:	**Leaves, flowers and fruit**
Predators:	**Harpy eagles, ocelots and jaguars**
Curiosity:	**Best friends with moths**

Once a week or so, a sloth slowly descends the tree.

It's time for a **toilet break**. But it's not that simple.

It's a risky business…

… **predators** are about.

The sloth poos and pees at the foot of its tree.

The **moths in its fur** jump on the pile of dung and **lay their eggs** in it.

The eggs hatch into **larvae**, which feed on the dung. The larvae change into **adult moths** and fly up into the tree.

The moths find a sloth to call home and **snuggle into its fur**. It's a strange **friendship**, but it works!

Bearded Vulture

A bone-breaking bird of prey

Bearded vultures are very large birds, with a big appetite for bones. Also known as lammergeiers, these scavengers have strong acid in their stomachs that quickly dissolves bone.

Bearded vultures mostly eat carrion (the remains of dead animals), especially bones, but they also attack living animals. They have been known to drop tortoises from a great height, so their tough shells break open.

The vultures **carry bones** up into the air...

... then drop them onto rocks below.

The **bones break** and the birds can eat the juicy **marrow** inside.

Bearded vultures are the only vertebrates that eat a diet that is made up, mostly, of bones. If they have too many bones to eat in one go, they store the spares in a cave or in their nest.

Bones are hard, but they contain soft marrow. It is fatty and full of goodness.

These birds prefer big bones to small ones, and old bones to fresh ones!

Fact File

Latin name: **Gypaetus barbatus**
Origin: **Europe, Africa, Asia**
Habitat: **Mountains and high grasslands**
Size: **90-120 cm**
Lifespan: **About 20 years**
Diet: **Carrion, especially bones**
Predators: **Birds of prey**
Curiosity: **Eats bones**

Goblin Shark

A long-snouted snapping shark

There are many curious creatures living in the deep oceans, where few humans have ever travelled. This is the world's most mysterious habitat, and home to the ghostly goblin shark.

Young sharks are called pups. Goblin shark pups are a ghostly white, but they turn pinkish as they get older.

Goblin sharks have freaky jaws that can shoot forward of its mouth – all the better for snapping up a passing fish in the blink of an eye.

Fact File

Latin name:	**Mitsukurina owstoni**
Origin:	**Atlantic, Pacific and Indian Oceans**
Habitat:	**Deep coastal waters**
Size:	**Up to 5.5 m in length**
Lifespan:	**Unknown**
Diet:	**Fish, squid and shellfish**
Predators:	**Unknown**
Curiosity:	**Toothy jaws that move forwards**

The jaws are lined with three rows of about 100 teeth. Some teeth are long and pointed, for grabbing and slicing flesh, while others are flatter for crushing shellfish.

The shark's long snout can sense the electricity produced by a fish's muscles as it swims.

The goblin shark **senses** its prey.

It **shoots out** its jaws.

Caught it!

Common Chameleon

The swivel-eyed lizard with fly-nabbing skills

Chameleons are famous for their ability to change colour, but that's not their only curious skill. These lizards can also grab their favourite food while barely moving a muscle, thanks to a super-speedy tongue.

Chameleons can swivel each eye independently of the other. It's a neat trick that allows them to look in different directions at the same time.

Chameleons can change the colour of their skin in seconds. They usually have green or brown skin, but become more colourful when they are communicating with each other.

Some chameleons can catapult their tongue towards their prey at a jaw-dropping five metres per second.

Most chameleons live in tropical parts of Africa. This is the common chameleon, the only species that is found in Europe.

Fact File

Latin name:	**Chamaeleo chamaeleon**
Origin:	**Europe, North Africa and the Middle East**
Habitat:	**Grasslands, forests and farmland**
Size:	**7-16 cm**
Lifespan:	**3 years**
Diet:	**Insects and fruit**
Predators:	**Snakes, birds and rats**
Curiosity:	**Colour-changing skill and speedy, sticky tongue**

A chameleon waits patiently for **insects or other bugs** to stroll by.

Its tongue is getting ready...

At **lightning speed**, the **long tongue** shoots out of the lizard's mouth, towards its victim.

The **sticky tip** of the tongue **glues to the bug**, and the lizard whips its tongue back into its mouth for a lip-smacking snack.

Blue Sea Slug

Stunning sailors with a toxic twist

Nudibranchs, or sea slugs, are some of the most gorgeous creatures alive. Their stunning colours are used for camouflage, or to warn predators that they are carrying venomous stingers. The blue sea slug is one of the world's most beautiful – but toxic – creatures.

Fact File

Latin name:	**Glaucus atlanticus**
Origin:	**Pacific, Atlantic and Indian Oceans**
Habitat:	**Open ocean**
Size:	**3 cm long**
Lifespan:	**Up to one year**
Diet:	**Hydrozoans**
Predators:	**Unknown**
Curiosity:	**Stunning beauty that steals stingers**

Nudibranchs are molluscs – a large group of animals that includes shellfish, octopuses, slugs and snails.

Seen from above, a blue sea slug is camouflaged against the blue water. It floats upside-down, or swims in pursuit of its favourite food, the Portuguese man o' war.

The slug's body is called a foot. On its head are rhinophores – special sense organs. Tentacle-like growths, called cerrata, grow out from the slug's body.

The blue sea slug is also known as the blue glaucus, blue dragon, blue angel and sea swallow.

The Portuguese man o'war is a type of venomous sea animal called a hydrozoan. Hydrozoans are similar to jellyfish and sea anemones.

The long tentacles of a man o' war trail below the water. Each tentacle holds polyps that have deadly stingers. A blue sea slug is immune to their venom.

The sea slug spots a man o' war.

It uses its **tiny teeth** to grasp the tentacles and eat the **polyps**.

The polyps' **stingers are absorbed** into the sea slug's body.

If **threatened**, the sea slug will unleash the stored venom.

Potoo

A spooky bird that thinks it's a tree

Northern potoos are peculiar birds that hunt at night, have huge eyes and sound like ghosts. Their feathers are flecked with browns, greys and creams. The colours and patterns create an awesome camouflage that turns this bird into a living, breathing bit of tree.

Fact File

Latin name:	**Nyctibius jamaicensis**
Origin:	**Central America**
Habitat:	**Tropical forests**
Size:	**Up to 44 cm long**
Lifespan:	**Unknown**
Diet:	**Insects and small birds**
Predators:	**Monkeys and birds of prey**
Curiosity:	**Looks like part of a tree**

When a potoo opens its eyes, enormous golden orbs are revealed.

A potoo chick is white and fluffy and will emit a buzzing sound when hungry.

Potoos perch, stock-still, on the tip of a broken branch. They can sit like that for hours, without moving a muscle.

A potoo spends the day asleep. As dusk falls, it prepares to hunt.

Perfectly camouflaged and still, the potoo is invisible to predators and prey.

Even the **chick is camouflaged** to blend into the tree.

Its **eyes are shut**, but the potoo can still **detect movement**.

As soon as **prey nears**, the potoo is ready to fly into the air to grab its victim.

Back on its perch, the potoo keeps an eye (or two) open for **hungry monkeys**.

Capybara

The giant rodent that barks like a dog

Capybaras love water. They lounge in ponds, swamps and rivers, where they browse on plants and swim with webbed feet. A capybara can hold its breath underwater for five minutes at a time, to hide from predators.

Rodents are everywhere – there are about 2,400 species of rodent in the world. Most of them are small mammals, with a furry body, four legs, a long tail and big front teeth for gnawing plants.

Basking in the heat of the day, a capybara makes the ideal perch for birds. In return, the birds snap up bugs and pests that bite or sting the rodent's soft skin.

Capybaras live in family groups. They all share the job of looking out for deadly snakes, jaguars and caimans.

Fact File

Latin name: **Hydrochoerus hydrochaeris**

Origin: **South America**

Habitat: **Wetlands and flooded grasslands**

Size: **1-1.3 m long**

Lifespan: **6-10 years**

Diet: **Aquatic plants and grasses**

Predators: **Anacondas, jaguars and caimans**

Curiosity: **World's largest rodent**

Capybaras are the biggest rodents in the world, but they are not the biggest rodents to ever live. Long ago, there were mega-rodents the size of bulls.

A capybara's **eyes and nose** are on **top of its head**.

This means it can wallow in the water while still breathing and **watching for predators...**

... or enjoy a **snooze in the sun**.

All females in the family group will help to look after baby capybaras (called pups).

Surinam Toad

A somersaulting toad that keeps its babies safe

Life as a small, soft and squishy tadpole can be dangerous – a juicy tadpole makes a tasty snack for fish. Female Surinam toads have an incredible way of protecting their growing young – they grow them in their skin!

Fact File

Latin name: **Pipa pipa**
Origin: **South America and Trinidad**
Habitat: **Ponds and swamps**
Size: **10-17 cm long**
Lifespan: **Up to 7 years**
Diet: **Worms, insects, crustaceans and small fish**
Predators: **Unknown**
Curiosity: **Grows its babies in its skin**

They are well camouflaged; their flat, grey-brown bodies look like dead leaves.

Surinam toads use their super-sensitive fingertips and special sense organs to detect movement in the water.

Surinam toads don't have teeth or a tongue. They suck in mouthfuls of water and swallow the small animals that float in it. They also use their nimble fingers to push prey into their mouths.

Most frogs and toads live as tadpoles in water, before moving onto land as adults. These toads, however, spend most of their lives in water.

Male toads **call to females** underwater.

The male hugs the female and they **somersault in the water**.

The female **releases about 100 eggs,** which the male fertilizes.

He pushes the eggs **onto the female's back**.

The eggs sink into her skin, and a **flap of skin** grows over the top of each one.

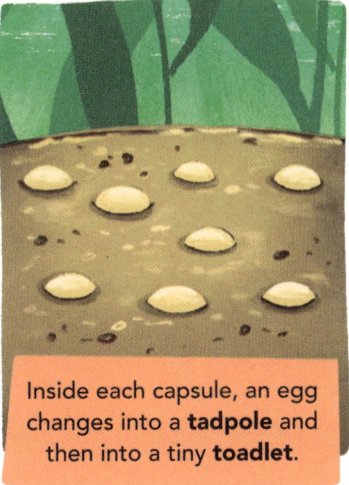

Inside each capsule, an egg changes into a **tadpole** and then into a tiny **toadlet**.

After several months, the toadlets **break out** of their mother's skin.

Peacock Spider

A rare beauty with dance moves to die for

A male peacock spider has one job to do: he must persuade a female to mate with him. He dresses to impress in splendid colours and shows her his very best dance moves. It is a do-or-die mission: if he fails, she may eat him!

A male peacock spider spies a female.

It's time to do the **fan dance...**

He waves **two hairy legs** in the air...

Then waits with his abdomen raised.

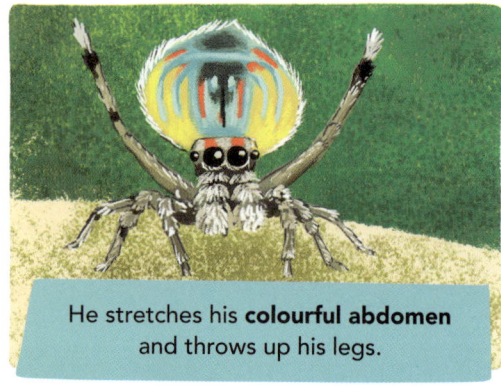

He stretches his **colourful abdomen** and throws up his legs.

The abdomen bobs up and down.

He **shuffles** from side to side.

He waves, **shakes** and **shimmers**...

The female watches, curious...

... he moves closer...

... and **bops her** on the head. He survived!

Fact File

Latin name: **Maratus volans**
Origin: **Australia**
Habitat: **A wide range of habitats**
Size: **Less than 5 mm long**
Lifespan: **Unknown**
Diet: **Insects and other spiders**
Predators: **Insects and other spiders**
Curiosity: **Gorgeous colours and dazzling dances**

There are more than 100 types, or species, of peacock spider. Each species has its own special pattern of colours.

Peacock spiders have eight eyes and incredible eyesight – they can see colours that are invisible to humans.

They leap to escape predators, jumping 40 times their own body length to reach safety.

A male taps his feet, waves his legs around and holds up his multi-coloured abdomen. The bold colours shimmer and shine as he dances.

Pink Fairy Armadillo

An armour-plated sand-swimmer

Beneath the sandy plains of Argentina, mysterious pink fairy armadillos burrow in the dark. Rarely seen, these tiny animals hide in their underground homes, coming out only at night. Very little is known about these curious creatures.

Fact File

Latin name:	**Chlamyphorus truncatus**
Origin:	**South America**
Habitat:	**Grasslands, scrubland and deserts**
Size:	**Up to 13 cm long**
Lifespan:	**Up to 4 years**
Diet:	**Insects**
Predators:	**Pet cats and dogs**
Curiosity:	**Shy and mysterious**

Pet cats hide, waiting for a chance to grab a tasty meal!

The pink fairy armadillo can blush: it flushes pink when it is warm, but turns pale when it is cold.

Pink fairy armadillos dig shallow burrows.

They use their flat bottoms to **push and flatten the soil** or sand behind them.

They spend most of their lives **underground**.

If they encounter an obstacle, they **burrow around**…

… or come to the surface.

It spends most of its life in its burrow, gobbling up ants. When it rains, armadillos race out of their shallow burrows, before they fill up with water. They hate getting wet!

Its soft, furry body gets some protection from an armoured shell made of bony plates, covered with a thick skin.

Huge claws are essential tools for a burrower. Scrabbling away at the sand, a pink fairy armadillo can disappear in seconds.

The armadillos are sometimes spotted by roads **after heavy rainfall**.

They need to move quickly away from the **dangerous speeding cars**...

... and **burrow to safety**.

Hoatzin

The stinky dinosaur bird

With their clawed wings, hoatzins have been compared to the first birds that evolved from dinosaurs. They have also been compared to cows, because of the way they digest their food!

Fact File

Latin name: **Opisthocomus hoazin**

Origin: **South America**

Habitat: **Forest trees by rivers**

Size: **Up to 65 cm in length**

Lifespan: **10 years or more**

Diet: **Leaves**

Predators: **Humans, birds, monkeys and snakes**

Curiosity: **Claws on wings**

When hoatzins are disturbed, they crash through the forest, making loud raspy calls. These sociable birds chat to each other with calls of 'ohh' and 'oww'!

These big birds eat a diet of leaves, which are very tough to digest. The leaves sit in a special pouch, called a crop. Bacteria break the leaves down into a green mush. This is similar to what happens in a cow's stomach.

As the bacteria get to work, they make stinky gases. The birds smell like cow manure and are often known as 'stink birds'.

Hoatzins **nest in trees above water**.

When danger looms, the **chick drops into the water below** to escape.

Splash!

When it's **safe**, the chick **swims back** to the tree.

It clambers up to the nest, using **two strong claws on each wing**.

The chick will lose these claws as it gets older.

Hagfish

A slippery, slimy scavenger

Hagfish look revolting, but these fish-like creatures are part of nature's clean-up crew. They have changed very little in the last 300 million years, and slime may be the secret of their success.

Hagfish **sniff out** a dead whale on the seabed, and tuck in.

A hungry fish approaches.

The hagfish looks tasty to the fish!

Under attack, the hagfish's **slime glands** get to work.

They release **tiny threads of slime** that unravel when they come into contact with water.

The **threads grow 10,000 times** their original size as they turn to slime.

The soft slime **clogs up a fish's gills**. It swims away leaving the hagfish to eat in peace.

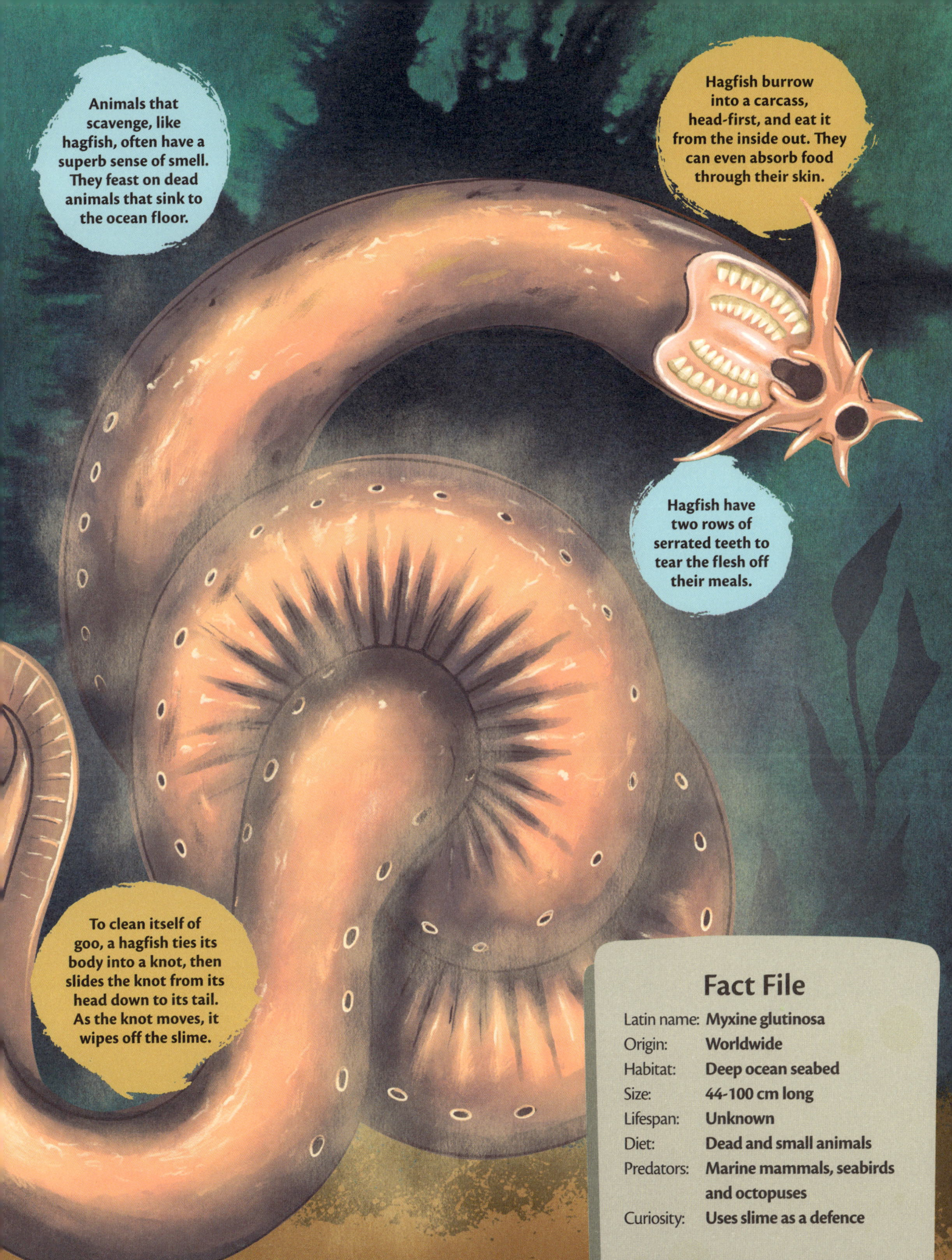

Animals that scavenge, like hagfish, often have a superb sense of smell. They feast on dead animals that sink to the ocean floor.

Hagfish burrow into a carcass, head-first, and eat it from the inside out. They can even absorb food through their skin.

Hagfish have two rows of serrated teeth to tear the flesh off their meals.

To clean itself of goo, a hagfish ties its body into a knot, then slides the knot from its head down to its tail. As the knot moves, it wipes off the slime.

Fact File

Latin name: **Myxine glutinosa**
Origin: **Worldwide**
Habitat: **Deep ocean seabed**
Size: **44-100 cm long**
Lifespan: **Unknown**
Diet: **Dead and small animals**
Predators: **Marine mammals, seabirds and octopuses**
Curiosity: **Uses slime as a defence**

Duck-Billed Platypus

A small egg-laying mammal – it's one of a kind!

Most mammals give birth and feed their young with milk, but the duck-billed platypus is special. It is one of just five unusual mammals that lay eggs, and the other four look nothing like it! In fact, this is one of the strangest animals to ever live.

A platypus walks like a reptile, has a beak like a duck, senses electricity like a shark, has a tail like a beaver, swims like an otter and lays eggs like a bird.

The thick, velvety fur is waterproof and the platypus's front feet are webbed.

Males have a poisonous spur on each back leg. They use the spur to deliver a dose of poison when fighting with other males.

A **female platypus** digs a **burrow** more than 10 m long. She **lays 2-3 eggs** in it.

She stays with her eggs for 10 days, **keeping them warm** until they are **ready to hatch**.

Before leaving the nest, the young **puggles feed on her milk** for about four months

As it hunts underwater, the platypus sweeps its head from side to side. It uses the information from its sensitive beak to create a 'map' of its surroundings and find its prey.

The flat beak, or bill, is covered with a layer of super-sensitive skin. It can detect the electrical signals made in the muscles of other animals.

Fact File

Latin name: **Ornithorhynchus anatinus**
Origin: **Australia and Tasmania**
Habitat: **Rivers, lagoons and streams**
Size: **40-60 cm in length**
Lifespan: **Up to 12 years**
Diet: **Insect larvae, crustaceans, snails, fish and worms**
Predators: **Foxes, dogs, cats, snakes and birds of prey**
Curiosity: **Egg-laying mammal with a toxic spur**

Fig Wasp

Friendly, fruit-loving wasps that can't live without a fruit - the fig!

There are about 900 species of teeny-tiny fig wasps in the world. They have evolved, over millions of years, to form a very special friendship with about 900 types of fig tree.

Fact File

Latin name: **Agaonidae**

Origin: **Tropical areas worldwide**

Habitat: **Woodlands, forests and gardens**

Size: **Up to 5 mm long**

Lifespan: **Often just a few days**

Diet: **Fig trees**

Predators: **Insects and birds**

Curiosity: **Fig friends**

Figs are not like most fruits. They have tiny flowers growing inside them.

Male fig wasps spend their whole lives inside a fig.

Fig trees need fig wasps to pollinate their flowers, so seeds can grow.

Each species of fig wasp pollinates just one type of fig tree.

Adult fig wasps may live for just one or two days.

A **female fig wasp** finds a fig and **burrows** into it.

She may **lose her wings and antennae** as she burrows.

She keeps burrowing.

She reaches tiny flowers, which grow inside the fig, and **lays her eggs**.

The wasp **pollinates the flowers** with pollen she has brought from another tree.

Exhausted, she dies.

Meanwhile, the eggs she laid begin to **develop**.

Blind, **wingless males hatch** out first.

They fertilise the females while they are **still inside their eggs**.

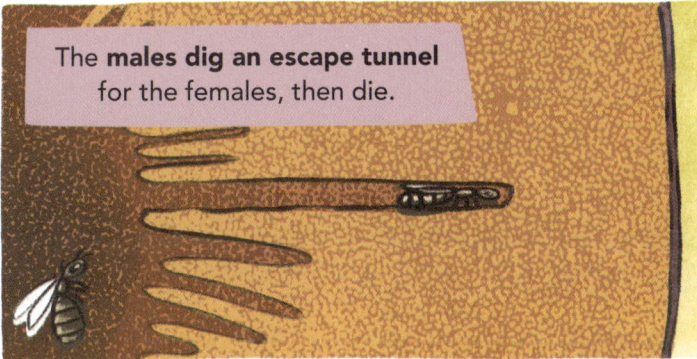

The **males dig an escape tunnel** for the females, then die.

The **females hatch** from their eggs…

… and **collect pollen** from flowers inside the fig.

They leave through the escape tunnel and find another fig tree to **start the life cycle again**.

Barreleye Fish

A deep-sea fish with remarkable eyes

Little sunlight passes through the ocean, and at depths of 600 metres or more, it becomes a gloomy, twilight world. The barreleye fish has a bizarre way of seeing in near-total darkness. It has enormous eyes and can see through its own forehead, which is transparent.

Fact File

Latin name:	**Macropinna microstoma**
Origin:	**Pacific Ocean**
Habitat:	**Deep waters**
Size:	**Up to 15 cm in length**
Lifespan:	**Unknown**
Diet:	**Small sea creatures**
Predators:	**Unknown**
Curiosity:	**Transparent (see-through) head**

Life is hard in the deep sea. There is **not much food** to find.

The barreleye fish is always on the lookout...

... for animals to eat, such as jellyfish.

It will eat **crustaceans** that are trapped in the **tentacles** of siphonophores.

A siphonophore is a colony of tiny animals that are related to jellyfish.

The transparent forehead is full of fluid to protect the fish's eyes and brain.

The eyes point upwards and forwards so they can spot prey swimming above.

Where a fish's eyes would normally be, there are two organs for sensing smell.

The fish has large, flat fins that it uses to swiftly change direction, or to remain motionless in the water while it is looking for prey.

Before humans arrived on New Zealand, kākāpōs were common birds across the country. They were hunted by humans, and by the mammals they brought to New Zealand.

By 1995, there were just 51 kākāpōs left alive. They were moved to new island homes, where there were no predators. There are now about 250 kākāpōs.

Now they only survive on three small islands, where they are protected.

Kākāpō

The owl-parrot that forgot to fly

The quirky kākāpō is one of nature's rarest curiosities. These are the world's largest parrots, but they cannot fly and they are nocturnal (active at night). These precious birds nearly died out, but they have been brought back from the brink of extinction.

Fact File

Latin name: **Strigops habroptilus**
Origin: **New Zealand islands**
Habitat: **Forests**
Size: **65 cm in length**
Lifespan: **45-60 years**
Diet: **Plants and fungi**
Predators: **Rats, cats, stoats and birds of prey**
Curiosity: **Nocturnal, flightless parrot**

Big and bulky, kākāpōs browse for food on the ground or climb trees to reach fruit.

These are noisy birds! They boom loudly and make 'ching' or 'skrark' calls.

When a kākāpō is scared, it freezes and waits for the danger to pass. Its green plumage helps the parrot to stay hidden in the forest undergrowth.

The fruit of the tall **rimu tree** is a favourite **treat for kākāpōs**.

The trees only **bear fruit every few years**. Kākāpōs choose this time to mate.

Kākāpōs use their wings and **clawed feet to climb** the rimu trees...

... where they enjoy a **well-deserved treat**.

Colugo
Nature's hang gliders

Dusk is falling in the rainforest, and the sounds of birds, frogs and monkeys echo in the gloom. There is a sudden movement in the canopy. A furry colugo has leapt from a branch and has taken to the air, silently gliding between trees.

Fact File

Latin name:	**Galeopterus variegatus**
Origin:	**Southeast Asia**
Habitat:	**Rainforests**
Size:	**Up to 70 cm long, including the tail**
Lifespan:	**Up to 17 years**
Diet:	**Plants**
Predators:	**Humans and birds of prey**
Curiosity:	**Glides great distances**

A colugo's large eyes face forwards. This helps it to judge distances, even after sunset when it is most active.

The size of a pet cat, a colugo has four legs and strong claws for clasping to a tree trunk. Colugos spend their whole lives in trees.

Colugos are often called flying lemurs, but they are not lemurs and they don't fly. They glide!

Flaps of skin stretch from the animal's neck to its fingertips, the tips of its toes and tail. This is called a patagium.

One baby is born at a time. Its mother carries it on her belly, or wraps it in her hammock-like patagium.

A **colugo** gets ready to **leap** from its tree branch.

It stretches out its patagium as it leaps through the air. It takes on the shape of a kite as it glides.

Landed safely! Gliding can put colugos at risk of being **attacked by eagles**.

A colugo can easily **travel 70 m between trees** in one glide, while barely losing height. The record is 136 m.

Did You Spot?

Great Black Hawk

Shares a habitat with... a hoatzin

Latin Name:
Buteogallus urubitinga

Size:
Approx 60 cm in length

Fun Fact:
Hunts in various ways and
eats almost everything

Sophonophore

Shares a habitat with... a barreleye fish

Latin Name:
Siphonophorae

Size:
The giant siphonophore can grow to 45 metres long –
longer than a blue whale!

Fun Fact:
There are 175 species of siphonophorae

Sloth Moth

Shares a habitat with... a three-toed sloth

Latin Name:
Cryptoses choloepi

Size:
Up to 10 mm

Fun Fact:
Sucks up moisture from a sloth's eyes and skin

Cattle Tyrant

Shares a habitat with... a capybara

Latin Name:
Machetornis rixosa

Size:
Up to 20 cm in length

Fun Fact:
May steal the nests of other birds

Egyptian Grasshopper

Shares a habitat with... a common chameleon

Latin Name:
Anacridium aegyptium

Size:
Up to 6 cm in length

Fun Fact:
Parts of the back legs have black stripes

Portuguese Man O' War

Shares a habitat with... a blue sea slug

Latin Name:
Physalia physalis

Size:
Up to 30 cm long in its body; 50 metre long tentacles

Fun Fact:
When threatened, it can deflate its body to submerge quickly

Belted Cardinalfish

Shares a habitat with... a bobbit worm

Latin Name:
Apogon townsendi

Size:
Up to 6 cm in length

Fun Fact:
Hides inbetween the spines of an urchin

Deep Sea Red Crab

Shares a habitat with... a goblin shark

Latin Name:
Chaceon quinquidens

Size:
Up to 175 cm in width

Fun Fact:
Has red eyes as well as a red body

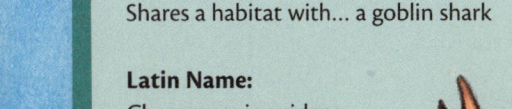

Smallscale Slickhead

Shares a habitat with... an anglerfish

Latin Name:
Alepocephalus australis

Size:
Up to 69 cm in length

Fun Fact:
Little is known about this deepsea fish, but it may be able to remove its entire stomach to empty it, before swallowing its contents again

Atlantic Wreckfish

Shares a habitat with... a hagfish

Latin Name:
Polyprion americanus

Size:
Up to 1.8 metres in length

Fun Fact:
So named because it is often found hunting around shipwrecks

Glossary

algae: tiny, plant-like things that often live in water.

amphibian: an animal – such as a frog, toad or newt – that has smooth, damp skin.

bacteria (plural of bacterium): a group of very tiny, very simple living things, found in almost all environments.

bait: something that looks like food, used to attract or lure prey.

browse: to reach into trees or shrubs to find and feed on vegetation.

camouflage: body patterns, colours and shapes that help to hide animals in their natural environment.

canopy: the thick, green tops of trees that stretches over the lower layers of a rainforest.

carcass: the dead body of an animal.

chick: a young or newly-hatched bird.

crustacean: an animal without a backbone that lives in the sea. Lobsters and crabs are examples of crustaceans.

display: the behaviour of an animal that is trying to attract a mate.

dissolve: when a solid mixes completely with a liquid, forming a solution.

endangered: at risk of becoming extinct (dying out).

evolution: a slow process of change that enables living things to adapt to the world around them. During evolution, new species develop, replacing those that have become extinct.

extinction: a species of plant or animal is extinct if there are no more living examples of it on the planet.

fertilization: the process that enables a male and female cell to join together, to form a single egg.

fresh water: water that is not salty, found (for example) in rivers and streams.

fungi (plural of fungus): a group or organisms that includes yeast, moulds, truffles, toadstools and mushrooms.

gills: body parts used to take in oxygen from water, so that an animal can breathe underwater.

grassland: a large, open area of land where grasses are the main type of vegetation.

habitat: the natural home where a plant, animal or other organism is suited to living.

insect: an animal – such as an ant or butterfly – that has six legs and three parts to its body.

lagoon: an area of salty water, separated from the sea by a sandbank or a coral reef.

larvae (plural of larva): a larva, or grub, is the early form of an animal before it changes into its adult form.

lure: a body part on an animal used to attract prey.

mammal: an animal that breathes air, grows hair or fur, and which feeds on its mother's milk when it is very young.

marrow: a soft, fatty substance found at the centre of bones.

metamorphosis: the stages, in the life cycle of an animal, through which it changes from its young, juvenile form into its adult form.

mollusc: an invertebrate animal, with no backbone, such as a slug, snail, mussel or octopus.

mottled: marked with spots or patches of colour.

patagium: a flap or fold of skin between the front and rear limbs of some mammals.

plumage: a bird's feathers, especially those that are long or brightly coloured.

pollen: a dust-like substance that flowers make to produce seeds.

pollinate: when plants use animals to spread their pollen from one flower to another.

polyp: an animal that has a tube-shaped body, with a ring of tentacles around its mouth.

predator: an animal that hunts and feeds on other animals.

prey: an animal that is hunted and eaten by another animal.

puggle: a baby platypus.

rainforest: thick regions of forest, often found in tropical parts of the world, where there is frequent rainfall.

reptile: an animal – such as a lizard, snake or tortoise – that has scaly skin. Most reptiles lay eggs that have soft shells.

scavenger: an animal that feeds on the remains of dead animals.

scrubland: an area of land where low shrubs, grasses and herbs are the main types of vegetation.

serrated: objects that have a saw-like or jagged edge.

species: a group of living things that have similar features or characteristics.

tadpole: the young, legless, larval stage of an amphibian – such as a frog, toad, newt or salamander – that breathes through gills until it grows into its adult form.

venomous: describes a plant or animal that can poison another creature.

vertebrate: an animal that has a skeleton and a backbone, or spine.

wetland: a marshy area of land, flooded with water for all or part of the year.